Worth A Thousand Words

Scanning for the Real World

By Phillip W. Moffitt and Rick Smolan

ofoto™

"One ought, every day at least, to see a fine picture."
— Johann Wolfgang von Goethe

A Light Source Publication

Credits

Phillip W. Moffitt
Rick Smolan

Creative Director
Lori Barra Nason/TonBo designs

Writer
Bernard Ohanian

Managing Editor
Amanda Jones

Photography
Doug Menuez/REPORTÁGE

Graphic Production
Jan Martí/Command Z
David Bender

Art Coordinator
Liz Rico

Copyeditor
Christine Shuken

Special thanks to
Sandra Koenig, Chuck Routhier,
Erin Murphy, Laura Tarrish,
Lisa Nelson, Laura Civiello,
Blake Sorrell, Rachel Gotbaum,
Barbara Troast

Cover Photograph
David Magnusson

Printed by
Pacific Lithograph

Copyright © 1991
by Light Source Computer Images, Inc.
Published by Light Source Computer Images, Inc.
17 East Sir Frances Drake Blvd.,
Larkspur, California 94939

All rights reserved, including the
right of reproduction in whole or in
part in any form.
Library of Congress Catalog Card Number:
91-062160
ISBN 0-9630008-0-2
Designed by Lori B. Nason, TonBo designs
Printed in the United States of America
10 9 8 7 6 5 4 3 2
Second Edition, 1992

The Light Source logo is a registered trademark of Light Source and
Light Source and Ofoto are trademarks of Light Source Computer Images, Inc.
Apple and Macintosh are registered trademarks of Apple Computer, Inc.

Contents

Introduction

The Logic of Images

5

Chapter One

Scanning for Profit

Esquire
Prudential Realty
Phelan's
Barking Dog Construction
The Avenue Grill

8

Chapter Two

An Office Away from the Office

The Cerre family
Sharon Garrett

22

Chapter Three

The World of Publishing

In Health
Elliott Erwitt
Magic Theater

28

Chapter Four

A Visual Education

Davidson School
Cynthia Liu

36

Chapter Five

Designing Like a Pro

Pentagram Designs
Carbone Smolan
The Access Books

44

Chapter Six

Pictures on File

REPORTÁGE
Acme Scenery
Nancy Hayes Casting

52

Chapter Seven

Printing: the Final Step

TechArt

63

The Logic of Images
by Phillip W. Moffitt

Since the earliest stages of human development, images have been an integral part of communication. As modern men and women we marvel at the drawings on the Lascaux caves in France or at the petroglyphs of Native Americans; we are moved by how powerfully they still convey information and feeling to us after so many millennia. Similarly, despite all our efforts to develop a rich written language in which words have the power to create "pictures" in our minds, it often remains true that — as the old Chinese proverb tells us — a picture is worth a thousand words. For instance, compare the following dictionary description of a kiss with the photograph beneath it:

> *"A touch or caress with the lips, often with some pressure and suction, as an act of desire or greeting."*

Despite all our efforts to develop a rich written language in which words have the power to create "pictures" in our minds, it often remains true that — as the old Chinese proverb tells us — a picture is worth a thousand words.

The average reader takes about twelve seconds to read, interpret and react to the written description, while the photograph delivers its message immediately. Additionally, the photo instantly conveys information about a particular kiss's type, purpose and mood; achieving the same effect verbally would have required many more words and involved even more time to comprehend. It's not that images are superior to words; it's that words and images are a natural pairing, which together can communicate more than either can alone — and in far less time.

In fact, the combination of visual and verbal communication offers numerous advantages over verbal methods alone: accessibility, efficiency, and, of course, the subliminal influence of aesthetics. Images are the extended arms, welcoming us into the full depth of a story that unfolds with well-chosen words. Images beckon to us, capture our attention more quickly than words and lead us efficiently where our true interests lie.

Yet while almost everyone is familiar with the evolution of written documents — from hand scribes to the Gutenberg press to the once ubiquitous electric typewriter and eventually to the desktop publishing equipment of today — relatively less attention has been paid to developing the capacity to incorporate images into books and documents. As a culture, we have by now become accustomed to multi-layers of stimulation — films with their rich sound tracks, live theatre with its masterful use of light, sound and special effects, and recording artists with videos to promote their musical work. But the world of the written word has not succeeded in incorporating images on a broad scale, primarily because the technology has been expensive and difficult to use.

As a culture, we have by now become accustomed to multi-layers of stimulation—films with their rich sound tracks, live theatre with its masterful use of light, sound and special effects, and recording artists with videos to promote their musical work.

But in the 1990s, the techniques for reproducing images will finally improve dramatically, thanks to a whole generation of new technologies. This revolutionary improvement will occur almost simultaneously for the professional publisher and the desktop publisher, leading to an explosion in the frequency and occasions in which we use images to communicate.

The key to this greater access to images is the use of a scanner, which digitizes images and therefore can reproduce them with remarkable accuracy. Until now, scanner hardware has lacked the software capable of fully utilizing the technology's potential while being simple enough for a non-expert to use. But with Ofoto™ software, a desktop publishing system can produce images that are as good as — and in many instances even better than — those resulting from traditional publishing methods. Ofoto™ works for all manner of texts, including magazines, newsletters, business letters and most other imaginable types of document, and is easy to learn and quick to use. The ability to create captivating, concise and informative documents, presentations and letters is now at hand.

The purpose of Worth A Thousand Words is to bring to life, in a clear, accessible way, the world of which we speak. Throughout this book, you'll find scanned images combined with text to show how individuals, families and businesses can use scanners to enhance the power of their work. Only a handful of our case studies highlight professional designers or publishers; on the contrary, nearly all of them feature people who, as recently as a year ago, could not have imagined combining images and text so easily. We hope you are intrigued and inspired by their work.

With new developments in scanning software technology, the ability to create captivating, concise and informative documents, presentations and letters is now at hand.

Chapter 1

Scanning for Profit

Saving time, saving money and looking sharp

If, as the old saying goes, the business of America is business, then there's an updated slogan for the 1990s: the business of business is communication. And because words, powerful as they may be, are often insufficient to illustrate a point in all its clarity and all its complexity, the scanner is becoming an indispensible addition to the phones, fax machines and computers that make up the office communication repertoire.

The desktop computer revolution of the 1980s signaled an extraordinary leap in the quality of text documents and put the power to create such documents directly into the hands of those writing the text. The desktop scanner revolution of the 1990s will signal an extraordinary leap in the relationship between text and images in business communication, and it will put the power to create text-image documents directly into the hands of those who understand intimately the message to be communicated. Because scanner technology has progressed to the point that anyone familiar with the use of a desktop computer can use a scanner effectively, executives and entrepreneurs with little or no background in design are illustrating business letters, incorporating photographs and drawings right alongside the text in proposals, creating brochures and catalogs, and generally giving all their business communications a professional, high-quality look. Magazines, newspapers, movies and television have long thrived on the marriage of words and images; now, finally, business communication is following suit.

There's an important caveat for every scanner user: tempting as it is, you can't simply scan — and use — every single beautiful image you see. Unless you create an image yourself by taking a photograph or drawing a piece of line art, the chances are good that the image will be copyrighted. As the name implies, copyrighting gives the image's owner — usually but not always the same person as the image's creator — the exclusive "right to copy" the work. The owner is empowered to give you permission to use the image, but without that permission — preferably in writing — you're in violation of the law if you use a copyrighted image. And even if you take a photograph yourself, it's standard practice to obtain permission from the photograph's subject before using it for commercial use.

In the pages that follow, you'll meet several people who have legally and effectively incorporated a scanner into their business lives: the associate publisher of a celebrated national magazine, who has made scanned images an integral part of her business letters, sales proposals, and client presentations; a high-powered realtor, who can use a scanner to show busy clients several properties at once; the president of a small but growing riding apparel business who can use scanned images to save time and money in the preparation of her thrice-yearly catalogs; the owner of a one-man construction company who uses scanned images to show clients his work; and the owners of a successful restaurant, who can use a scanner and copyright-free art to decorate their changing menus.

As varied as the scanner's business uses profiled throughout this book are, they only scratch the surface of the possible; for, over the next few years, the same creativity that characterizes American business at its best will undoubtedly lead to exciting new uses for a scanner. There are as many different uses for a scanner, after all, as there are kinds of businesses. And if the business of these businesses is communication, the essence of their communication must be an exciting combination of words and pictures.

TIPS for business use

❶ *By scanning your logo into your computer and storing it digitally, you can easily access it — inserting it wherever and whenever you want into your letters and other business communication. You have to scan an image into your computer only once to be able to use it an infinite number of times.*

❷ *Scanning your photo onto your business card helps prospective customers match a face with your name.*

❸ *Using scanned images in your business letters and sales proposals can effectively illustrate your product or service for your clients and prospective clients — especially if you customize the image where possible to show how your product can work for them.*

❹ *If you send out, as part of your sales or promotional material, photographs of clients or people not employed by or associated with your business, be sure to secure their written permission. The best and easiest way is to ask each person to sign a standard "model release."*

Visual Aids for Sales Proposals that Mean Business

In the highly competitive world of magazine advertising sales, every advantage can translate into hundreds of thousands of dollars. "Selling clients on a special advertising section of the magazine is made easier," says Esquire associate publisher Cheryl MacLachlan, "when I can actually show them ahead of time what the section might look like, and what their ads will look like within the section." MacLachlan uses a scanner to include visual aids, such as photographs, in prospecting letters to clients; later in the sales cycle, the scanner allows her to prepare a customized section mock-up for each potential advertiser, without incurring the expense of freelance artists. "The scanner is easy and fast enough that I can prepare the material myself; I don't have to work around anyone else's schedule, and it's much easier to customize each presentation. Additionally, because we're more efficient, we're able to increase the number of prospects we approach."

Esquire

Mr. Peter Butler
BBDO
1285 Avenue of the Americas
New York, NY 10019

Dear Peter,

Not wanting to subject you to a "new name from out of nowhere" — I'd like to take the opportunity to introduce myself via this letter. A quick background note:

Esquire is embarking on an exciting new project which is very well positioned to support your client's efforts in the new international campaign. Although I want to wait until we can meet in person to provide all the details — I can say that this project will be precedent-setting in terms of its scope. Not only will it allow you to reach the readers of Esquire, but we are able to deliver an additional universe of prospects as well.

One of the creative directors on this project is Rick Smolan, the man who created the enormously successful "Day In The Life" series. Pictured here are just two of the books from a series that included America, USSR, Japan, Australia and many more. I take the time to mention this, not because the new project is in any way a photojournalism book, but because the "Day In The Life" captured an emotion that proved very

...aints that you must work within during ...t we will be proposing something that

...etter — and I look forward to ...roject.

Esquire

Mr. Carlos Gutierrez
The Kellogg Company
1 Kellogg Square
Battle Creek, MI 49016

Dear Carlos,

I hope I haven't violated any rule of "coupon protocol" in substituting the word "breakthrough" for "great taste" — but I do feel optimistic that we may be on to something new and exciting here.

As a result of much input from John, we have made a significant breakthrough in the design of the project. Particularly in the area of out-of-pocket cost and cost-per-thousand. We have approached the program with some new thinking — which I believe will allow you to find it a much stronger proposition for Kellogg.

Additionally, I had a chance to speak to Charlie about the potential for a couponing program. Although I was only able to describe the program over the phone, Charlie saw a very good fit, editorial-wise. He wasn't convinced that couponing was the best way to go for a value-added program — but he was excited about an innovative idea he had to make this a strong opportunity. Charlie and I agreed to meet in person so that he could review the materials first-hand and see if his ideas did indeed have any application.

I would love to have a second meeting with you to show you the significant changes we have made in the program. I am trying to set up a trip to Illinois and Michigan for the week of April 28. Would you have any time on May 1? I will call on Tuesday, April 16, to see if we might schedule some time.

As always, thank you for your time and interest,

Cheryl MacLachlan

"My letters have a lot more punch when I illustrate my points graphically," says Esquire associate publisher Cheryl MacLachlan, who often scans in mocked-up ads or other visual aids to underscore her point. "It's almost as if they're transformed from letters into audio-visual presentations."

.....CLEAR CONNECTIONS.....CLEAR CONNECTIONS.....CLEAR CONNECTIONS.....

MacLachlan also uses a scanner to custom-prepare overhead presentations for potential clients. This mocked-up ad for a telephone company features a photo lent to MacLachlan by friends, so as to avoid copyright infringement or violation of privacy.

Practical Matters

Real Talk

Don't let long distances come between you and your family

Lorem ipsum dolor sit amet, consectetuer adipiscing elit, sed diam nonummy nibh euismod tincidunt ut dolor sit amet, consectetur

Lorem ipsum dolor sit amet, consectetuer adipiscing elit, sed diam nonummy nibh euismod tincidunt ut. Zorem ipsum dolor sit amet, consectetuer adipiscing elit, sed diam nonummy nibh euismod tincidunt ut lareet dolore magana aliquam erat volupat. Autem vel eum iriure dolor in hendrerit in vulputate velit esse molestie conseciptit lobortis nisl ut aliquip ex ea compmodo conseNam liver tempor cum soluta nobistis nisl ut aliquip ex ea compmodo consenam liver tempor cum soluta ut

aliquip ex ea compmodo consenam sed diam nonummy nibh euismod. Xorem ipsum dolor sit amet, consectetuer adipiscing elit, sed diam nonummy nibh euismod tincidunt ut lareet dolore magana aliquam. Autem vel eum iriure dolor in hendrerit in vulputate velit esse molestie consecipit lobortis nisl ut aliquip ex ea compmodo conse. Nam liver tempor cum soluta nobistis nisl ut aliquip ex ea compmodo conseNam liver tempor cum soluta

Stretching Your Long Distance Dollar

1. Lorem ipsum stamet, consecteturer adipscid-ing elit, des diam non-ummy nibh.

2. Duis autem vel eum possim sauum. Duis auterm vel euim iriruer dolor.

3. Lorem ipsulm dolor sit ament, docesttueer adminum.

4. Mazin placerte faccer possim saumsa. Duis autem vel ipsum lorem.

5. Lorem ipsum stamet, consecteturer adipscid-ing elit, des diam non-ummy nibh.

6. Duis autem vel eum possim sauum. Duis auterm vel euim iriruer dolor.

7. Lorem ipsulm dolor sit ament, docesttueer adminum.

8. Mazin placerte faccer possim saumsa. Duis autem vel ipsum lorem.

─────── *Esquire Special Promotion* ───────

THIS MARCH ESQUIRE READERS WILL DISCOVER AMERICA BY BICYCLE

ESQUIRE READERS ARE ACTIVE
- 35% Bicycle Regularly
- 32% Jog 3x or More a Week
- 45% Belong to a Health Club

ESQUIRE READERS TRAVEL
- 67% Take 5+ Domestic Trips
- 57% Take 3+ International Trips

ESQUIRE READERS ARE EDUCATED
- 85% Attended/Graduated College
- 36% Have Done Postgraduate Study
- 12% Have a Doctoral Degree

PHOTOS: DAVID MADISON

MacLachlan scanned color photos from a bicycle touring company to prepare a high-quality, black-and-white mock-up of a special advertising section on bicycle touring. Potential advertisers could see what the section might look like, and even how their ad — also scanned in by MacLachlan — would fit in to the section, visually and editorially.

MacLachlan's visually-aided sales campaign for the special advertising section on bicycling was a success, and the section ran in an issue of Esquire featuring the late Jim Morrison on the cover.

An Esquire Section Proposal

Helping Clients Picture the House of Their Dreams

In the rarefied world in which Brigitte Johns operates, time is indeed money. "My clients are busy people," says the Prudential real estate agent, "and I can't afford to waste their time by taking them to a house that isn't right for them. It's great to show them a picture of the house so that they can decide whether or not to go out in person." Johns can save time for her clients by scanning into her computer professional photos of dozens of homes, and creating a database in which homes are categorized by size, price and various amenities. Clients who visit her office can then look, on screen, at houses that fit their requirements. Johns also prints and sends out customized flyers — targeting clients with homes that may interest them, with personal messages attached. "I even send the flyers to out-of-town clients who are relocating," says Johns. "The pictures are an enormous help."

Realtor Brigitte Johns can print out, directly on her laser writer, custom flyers such as this one, with a special personalized message for potential buyers.

*O*ak and pine surround this handsome English Tudor home. Six bedrooms, soaring cathedral ceilings and a gourmet kitchen. Separate dressing room, detached office, huge wrap-around deck and pool. Offered at $989,000. For more information contact Brigitte Johns 415-331-0300 or 415-435-6207.

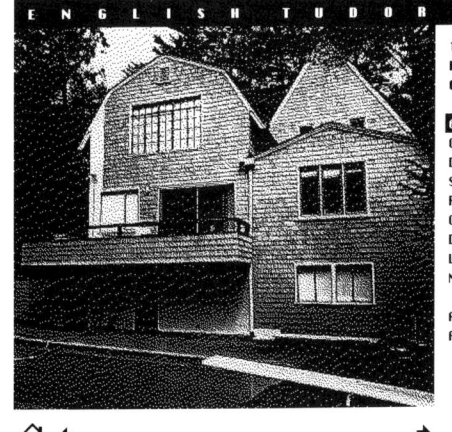

After scanning into her computer high-quality photos of dozens of properties, Johns can create a database (right) in which to quickly search for compatible homes.

Riding High with a Catalog of Text and Images

Patty Phelan, whose Sausalito, California, mail order business specializes in riding apparel, looks to control costs wherever she can. The time and cost of producing the company's thrice-yearly catalog are a major headache. The catalog is now produced using photostats to mock up numerous versions of the inside pages — each with several images and photographs — and the cover. The solution? A scanner, with which designer Bonnie Phippen of Phippen Design can vary the size, angle, cropping and location of each image on screen. "With a scanner I'll be able to review several versions of each page," says Phelan, "and we'll only have to size and shoot photostats once — when Bonnie and I have agreed on a page's final design." A scanner can also serve as an image library. Phelan's Equestrian catalog company offers popular merchandise in several successive catalogs, and designer Phippen can use a scanner to import into the computer, and store for re-use, photographs of these perennial items.

Catalog design requires a perfect fit, on every page, of several images and blocks of text. Phelan's catalog designer, Bonnie Phippen, can use a scanner to experiment with several designs of the same catalog page; once each image is scanned into the computer, it can be re-sized, cropped and moved around countless times, until Phippen and company president Patty Phelan find the right combination. The original image can then be stored digitially in the computer for re-use in subsequent catalogs.

Easy-to-Follow Plans from a Master Builder

Carpenter and remodeling specialist Simon Worrin of Barking Dog Builders knows first-hand the mutual frustration that occurs when a client expects one thing and the craftsman builds another. "I do everything I can to avoid surprises with a new client," says Worrin, "starting from a written proposal that outlines exactly what I plan to do." Experience, however, has taught Worrin that the most carefully worded proposal in the world can still be misinterpreted. That's why his proposals now include scanned-in images. "Along with sketches of the project at hand," he explains, "I include photographs of my previous work." Worrin's proposals also include photos of a variety of hardware accessories, scanned in directly from various catalogs, so that clients can choose exactly what they want. The final document, with words and images agreed upon by Worrin and his client, serves both as a record for the client and as a reference point for Worrin.

**68 Taylor Drive
Fairfax, CA 94930
(415)453-6854
License No. 2344562**

❶ Shelf Support

Mr. and Mrs. Ingrassia
146-14 243 Street
Garden City, Arizona
11702

PROPOSAL

BARKING DOG BUILDERS hereby proposes to furnish all materials and labor necessary to build and install cabinet/bookcase wall unit in living room on either side of existing French doors. Unit will be approximately 17' wide by 9' high. The depths of the cabinets and bookcases will be 21" and 12" respectively. The connecting display shelf above French doors will be 8' wide by 15' high by 8" deep.

Carcases will be constructed from 3/4" paint-grade shop birch plywood with 1/4" birch plywood back panels. Face-frames will be constructed from dressed 4/4" solid alder stock, biscuit-joined at joints, and nailed and glued to carcases. Cabinet doors and bookcase shelves will be constructed from 3/4" Medite.

Unit will conform to existing style of living room. This will be achieved by matching or milling facsimiles of existing mouldings, including crown-moulding, chair-rail, wainscoting astragel, base-cap, and baseboard.

Doors will correspond to wainscoting. This will be achieved by using ❸ concealed Blum cup-hinges and ❷ touch latches.

All shelves will be adjustable using ❶ brass-plated steel shelf supports or similar product/support. Shelves will be profiled along front edge to complement moulding details. Also, face-frames on bookcases will have a beaded edge.

Top shelves in bookcases and connecting display shelf to be illuminated by LyteTrim concealed incandescent lighting, switched inside cabinet.

Materials and labor to complete above: $_____.___

Terms: 10% at commencement, remainder on satisfactory completion.

Cabinet Bookcase
(left Partial view)

❷
Magnetic Touch Latch
Locks when pushed and when pushed against extends about 5/8" out.

❸ Concealed Hinges

All Our Dogs are Barking

A typical Barking Dog proposal includes a written description of the work Simon Worrin plans, along with a photograph of a previous project and photos of several examples — scanned in from the appropriate catalogs — of the hardware accessories to be used.

In Simon Worrin's pre-scanner days, his clients chose the latches and hinges they wanted by poring over photcopies of catalog pages, as pictured at right. Now Worrin can scan in various options from a number of catalogs, and put together an accessory sample sheet for clients. Working from the sheet of scanned photos — many of which Worrin has cropped and enlarged to highlight details and special features — Worrin's clients can directly compare items from different catalogs.

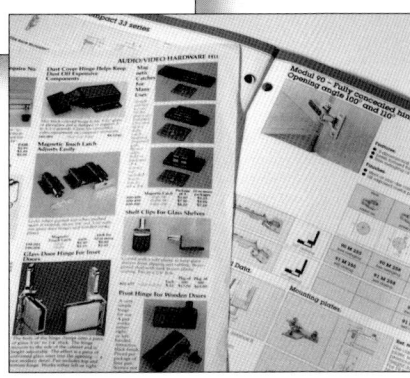

Clip Art in Menus Becomes a Couple's Meal Ticket

At The Avenue Grill in Mill Valley, California, hungry regulars have come to expect a new menu every month — which soon will be illustrated with scanned clip art images from the Dover series of copyright-free books. "We want our customers' dining experience to be a little bit different every time they come in," says Marni Leis, who owns and operates The Avenue Grill with her husband, Joe. "And changing the look of the menu helps keep us fresh as well." Each month, the Leises — neither of whom have any design experience — will scan the images they want into a template created by a designer, manipulating the size and resolution of the drawing as they see fit. They'll then print out the menu on their office laser printer, and take it to a copy shop to run off 200 copies on high-quality paper. The whole process should take no more than an hour, and the cost of these ever-changing menus will probably be no more than $20 per month.

WINE LIST

California Wines

WHITE

Lakespring '89 Sauvignon Blanc, Napa
4.25

Iron Horse '89 Fumé Blanc, Alexander Valley
23.00

Chalk Hill '88 Chardonnay, Sonoma
24.00/6.25

RED

Caymus '88 Zinfandel, Napa
21.00/5.50

Davis Bynum '87 Pinot Noir, Russian River
27.00

Guenoc '86 Cabernet, Lake County
22.00/5.75

SPARKLING

Scharffenberger Brut Non Vintage
25.00/5.25

THE AVENUE GRILL

MENU

Appetizers

Creamy provincial lobster bisque w/garlic toast
4.25/5.95

Velvet asparagus soup topped w/garlic brie
3.25/5.50

Fanny bay oysters on the half shell
7.50

Grand Bahamian tropical shrimp salad
sweet gulf shrimp, ripe avocado,
set upon mixed baby greens
9.25

Garden greens w/champagne dressing
4.95

Caesar salad
7.50

Avenue salad w/dried tomatoes,
two cheeses & tangy dressing
7.50

Tuscan toast w/roasted garlic,
parmigiano, & basil anchovy butter
5.95

THE AVENUE GRILL

MENU

Seafood Grill

Three pair of fresh icy cold oysters served
on the half shell
7.25

Plump belons w/a mignonette sauce
8.50

Steamed & chilled half
Dungeness crab
13.95

Grilled local swordfish
served w/shoestring potatoes
17.50

Fresh one & a half pound live Maine lobster,
steamed & grilled, w/New England clams, roasted new
potatoes & garlicked greens
38.95

Ahi tuna; lean & mean swordfish; & rarely
caught Lehi (deep water snapper)—all grilled in the
island tradition...w/Lomi Lomi salsa,
cucumber sauce & lemon butter sauce
18.75

THE AVENUE GRILL

With scanned images from the copyright-free Dover book, *Food and Drink—a pictorial archive from 19th-century sources,* Joe and Marni Leis of The Avenue Grill can create distinctive menus. "Scanning has become so easy," laughs Joe, "that even I can do it." The Grill's menus can be printed directly on a laser printer, and reproduced at a copy shop.

Wine labels, food images from clip art books, and The Avenue Grill logo itself can be scanned into the menu by Joe and Marni Leis.

An Office Away from the Office
And other uses for the home scanner

A major transformation is underway in American home life — a transformation that futurists and science buffs have been predicting for years, but one that just recently began in earnest. Electronics are becoming a significant presence in the home, most significantly in the home office, where an ever-increasing number of Americans are choosing to spend their days. The spare bedroom or family den has become a work center, with desktop computers, laser printers, fax machines, and now scanners.

And once the scanner is installed in the home as a business communication tool, the home office can also serve as a family fun center and educational resource. A scanner allows users to incorporate the power of images — the raw emotion inspired by photographs and other visual aids — into home-based communication. Imagine printing out drawings by your children as part of their letters to grandma. Imagine a birth announcement featuring a clear and crisp photograph of your newborn, right alongside the text that gives the details of the happy news. Imagine sending out invitations to a family reunion, with a sheet of directions featuring photographs of a few key landmarks and an easy-to-read map. Imagine creating name tags for everyone who attends that family reunion, with each name tag bearing a photograph of what the wearer looked like 20 years ago. Imagine a hobbyist having a digital record, on a home computer, of every single item in his or her collection of wine labels, stamps, postcards or political buttons. Imagine computerized recipe cards that you can categorize and call up quickly and easily, and which include a photograph of the finished meal.

For family fun, scanners turn scrapbooks into digitized multimedia keepsakes, as the family profiled in the following pages found after returning from a trip to the Spice Islands. Photos, letters, diary entries, even souvenirs become part of the package, as children and their parents create desktop-published "books" of their favorite experiences. What's more, these "books" — and the images used to create them — can be digitally stored and then revised countless times.

A scanner can also be simply and effectively used for more pragmatic household tasks. Having a garage sale? You can scan photgraphs of your most attractive items and quickly develop an inexpensive yet communicative flyer. Want to change the living room around? See how it will look before you do the heavy lifting, by scanning into your computer photographs of your furniture, then moving the scanned images around on your computer screen to experiment with the looks you want. Planning to remodel the kitchen? Scan in photos of your favorite kitchen elements from home design magazines, to see on screen how the parts fit together.

And, just as the desktop computer's use as a family budgeting device and as the linchpin of the home office make it a cost-effective purchase — one that goes beyond mere fun — so does the scanner pull its weight financially. As the example in the following pages illustrates, you can use a scanner to assemble a thorough record of your possessions for insurance purposes, with images combining with descriptive text to form a comprehensive report that is easy to follow, easy to file and easy to update.

TIPS for home use

❶ *You can turn your annual holiday letter into a family-produced booklet, with a family photo, drawings by your children and photographs of the year's most special moments.*

❷ *Images give added flavor to your party invitations: a photo of the birthday boy or girl, for instance, or an invitation to a silver anniversary party that features a photo of the happy couple on their wedding day.*

❸ *A commemorative booklet, following a once-in-a-lifetime event such as a first communion, bar mitzvah or college graduation, is easily prepared with the help of a scanner. It can feature photos of the special day, a guest list and comments from key participants. Copies are easily made and can be sent to party attendees as well as out-of-town friends and relatives who missed the fun.*

❹ *For safekeeping and insurance purposes, keep in your computer scanned copies of photos of your valuable possessions.*

One Family's Electronic Diary of the Trip of a Lifetime

Family vacation memories come in all forms: snapshots, souvenirs, diaries, postcards, a child's drawings, and the stories told and re-told for years. What if, Mike and Gina Cerre wondered upon their return from a month on the tiny Indonesian island of Banda, these elements could be combined, so that their two daughters, 9-year-old Lauren and 6-year-old Lee, could produce a book about their exotic vacation? Enter the family scanner. Once back in the United States, Mike and Gina typed into their home computer the text of their daughters' diaries. Among the text, the girls scanned into a page-layout program their choice of photos, postcards, maps, Indonesian stamps and money, and their own illustrations. Mike and Gina then printed the pages on their laser printer, and bound them into a "book" that the girls were able to share with their teachers and classmates, and that the family will treasure for years to come.

The cover of six-year-old Lee Cerre's diary features family snapshots, her own drawings, stamps — even, as the background, a piece of Indonesian fabric. The elements were scanned, and in many cases re-sized.

Lee invited her classmates to see her book with an invitation created by scanning various elements. The scanning of money is usually considered legal if it is reduced or enlarged by at least 20% and is reproduced in black and white.

The raw materials for the vacation books published by Lauren and Lee Cerre.

Recording Valuable Household Items the Easy Way

It's the homeowner's nightmare: a home theft, or fire, results in the loss of valued possessions. And an effort to recover the full value is greeted by an insurer's request: "Do you have photographs?" Most savvy homeowners do have photos of their most prized possessions: stacks of Polaroids, perhaps, with a corresponding list of descriptions, serial numbers and appraised values. Snapshots, however, are easily misplaced, and reproduce badly when photocopied. There's a better way, as Sharon Garrett discovered. Despite having no previous scanner experience, Garrett hooked a friend's scanner up to her home computer and laser writer, and within a few hours had used her snapshots to create a permanent record of her possessions — matching a detailed description to each image. The color photographs reproduced quite sharply on the black-and-white scanner; once her catalog of personal possessions was printed out, she sent one copy each to her insurance agent and attorney.

List and photographs of insured assets: lamps and vases

Insured:
Sharon Garrett
24 Maine Street
San Francisco, CA 90271

Underwriter:
John Wright, State Farm Insurance Co.

cc: John Wright, State Farm Insurance Co., Bernadine Marconi, attorney-at-law, Connie Garrett, estate executor

Fearing that a pile of snapshots of her most valued possessions would be easily lost or separated from other important information, Sharon Garrett used a scanner to create a combined text-image record of her home's most valuable items.

❶ Early Columbian water jug
Purchased by Sharon Garrett
from Max Booth Furnishings
10-12-90 for $3,500
Appraised by Lloyd's of London
1991 at $12,000

❷ 1920 Tiffany desk lamp
Purchased by Sharon Garrett
from Classique Art
5-17-90 for $4,000
Appraised by Lloyd's of London
1990 at $20,000

❸ 1800 Chinese celadon lamp
Purchased by Sharon Garrett
from Max Booth Furnishings
6-30-89 for $8,000
Appraised by Lloyd's of London
1991 at $20,000

❹ Pre-Columbian water jug
Purchased by Sharon Garrett
from Classique Art
9-4-84 for $800
Appraised by Lloyd's of London
1991 at $2,000

❺ Art Nouveau chandelier
Purchased by Sharon Garrett
from Max Booth Furnishings
5-16-90 for $6,000
Appraised by Lloyd's of London
1990 at $8,000

Receipts and appraisals attached

The World of Publishing

A new tool revolutionizes the printed page

Freedom of the press — according to a saying attributed to several people, including the American newspaperman A.J. Liebling and the Soviet revolutionary V.I. Lenin — belongs to he who owns the press. It's perhaps a cynical interpretation of one of our most cherished rights, but one with more than a grain of truth.

While it remains expensive to actually own a press, two major technological breakthroughs in the last 40 years have spread freedom of the press, by making it easier and cheaper for the lay person to print newspapers, brochures, catalogs, flyers — virtually anything. The first milestone, the development of the offset press system of printing — based on photo-composed "cold" type rather than "hot" or cast-metal type — meant that in the 1960s, anyone with a typewriter, a pot of glue, a ruler and a rudimentary sense of design could create "boards" or "flats" from which a print shop could produce a professional finished product. Of course, typeset copy still looked much better than its typewritten cousin, but in the new "cold" type era, typesetters could make corrections much more quickly and easily than before. Gone were the days in which type had to be set, slowly and laboriously, by trained operators of keyboards that created cast-metal type. Typesetting shops sprang up, as did offset print shops, and the United States saw an explosion of printed information, much of it literally produced by the boy or girl next door.

The second innovation came in the 1980s, with the growth of desktop publishing (also known as personal publishing, electronic publishing or computer-aided publishing). With the development of sophisticated page-layout programs for personal computers, every writer became a typesetter, and every office with a computer and a laser printer became a black-and-white print shop. The country experienced a new wave of printed information, again often produced by people with little or no formal training in the fields of design or printing.

Publishing, then, has truly moved from a specialization to a mass phenomenon. But one aspect of publishing has remained a specialization, in the hands of experts: the preparation and placement of images in computer-generated documents. Ever more sophisticated drawing programs have allowed desktop publishers to illustrate their documents by creating original art, but the insertion into a document of existent photographs or line art has remained the responsibility of the local print shop. The development of desktop scanners was to have changed that, by allowing users to transfer images directly into their computers and therefore into their computer-generated documents. But until now, scanning software has required users to have a working knowledge of terms like dots-per-inch, scaling factors, calibration, even densitometers. In addition, images have been almost impossible to reproduce without a degree of crookedness, and cropping, enlarging and reducing have presented major challenges.

That's where new developments in scanning software come in. Easy-to-operate selection controls allow the inexperienced user to focus simply on the desired size and shape of the final image; the rest of the settings — such as bit depth and dots-per-inch — are then set automatically. (Experienced users, of course, can still adjust the settings themselves to achieve sophisticated special effects.) Scanning now requires no more special knowledge of printing and design than does the operation of a camera or a copy machine. And while an award-winning magazine art director, a famed photographer and the general manager of a non-profit theatre company may have different needs, the same scanning software now works for all of them.

TIPS for publishing use

❶ *Set up a calibration for each paper stock and printer you might use when outputting your scanned images. Once you've set each calibration, you never have to do it again.*

❷ *As an occasional variation to the normal halftone screen that your scanner produces of a photo, use the software's mezzotint or dithering feature to give photographs a slightly tinted effect.*

❸ *If you plan to rotate or change the scale of an image, do so before placing the image into your page layout program. The results will be cleaner and clearer.*

❹ *Set up templates for any work you do on a recurring and regular basis — business letters, newsletters, monthly sales reports.*

❺ *If you're taking your publication to a service bureau or print shop for the first time, be sure to call ahead and ask what services they offer and in what form they prefer to get the document.*

Choices Made Easy at Award-Winning Magazines

The production of a magazine is the quintessential team effort, with writers, editors, designers and art directors pooling their talents. And for Jane Palecek, the art director of the award-winning magazines "Hippocrates" and "In Health", the ability to present design options to the other members of the publishing team quickly and efficiently is a crucial part of success. To prepare a photo essay, Palecek scans in the photographs she's chosen and then looks at several sizes and crops for each one — finally finding an attractive design that fits with the explanatory text necessary for each picture. Even before printing, she can see, on her computer screen, exactly what the photographs will look like at a variety of sizes and crops. She then prints the proposed design for the essay and presents it to the editors for approval, knowing full well that any adjustments they ask for are easily made by moving around, or re-sizing, the scanned images.

Jane Palecek of "Hippocrates" and "In Health" magazines has found that, with a scanner, she no longer needs to produce countless photostats to look at several sizes and crops for each photo in a photo essay. She can produce a draft of the essay for editorial approval, changing it quickly and easily if necessary.

PHOTOS: MAX AGUILERA-HELLWEG

A Noted Photographer Puts His Work in Order

Photographers thrive on creative control. By deciding where to point their lens, and by selecting their best images, they can choose what we see and don't see. Often, though, they see their creative control slip away when publishing books of their work — books that are often designed and laid out by someone with taste and style different from the photographer's. Elliott Erwitt, author of eight books and winner of numerous awards, has found a solution to the book design problem: a scanner. Erwitt can quickly and easily scan into his computer the photos he wants to use and can lay out his books himself — presenting the design to his publisher and saving time in the process. Throughout the world of publishing, authors are discovering that, increasingly, they can print directly from scanned images, without having to re-shoot photographs as photostats. Thanks to technological advances, the distinction between author and publisher is fast disappearing.

Cannes, France, 1975

Cannes, France, 1975

Famed photographer Elliott Erwitt can use his scanner to digitally store his work and to lay out the photos in book form in the exact order and style he'd like them to appear.

Bratsk, Siberia, U.S.S.R., 1967

New York City, U.S.A., 1977

New York City, U.S.A., 1953

New York City, U.S.A., 1974

With Pictures, Ads and Logos, the Playbill's the Thing

With one eye on the stage and the other fixed firmly on the bottom line, Mary Anne Cook of San Francisco's Magic Theatre is a typical general manager of a non-profit arts organization. Money is tight, and time is short. With publishing technology now more powerful and easier to use than ever, Cook has come to rely on it to keep costs in line without compromising on quality. When Cook first used a scanner, she realized that a star was born: "It's perfect," she says, "for creating the playbills for our seven annual productions." Each playbill — an attractive 8-page brochure featuring a history of the play and short biographies of the playwright and actors — includes photos of the actors, as well as ads Cook creates by scanning in logos. "We can change the size of ads each issue," she says, "and the photographs come out so crisply that our service bureau can print directly from our computer disk, without having to shoot photostats."

MAGIC THEATRE

Harvey Seifter, General Director

Presents

TEMPTATION
By **VACLAV HAVEL**
Directed by **HARVEY SEIFTER**

Tom Luce (L) and Bruce Williams (R). Photo: Bob Hsiang.

WHO'S WHO

VACLAV HAVEL, Playwright, was born in Prague in 1936. His first play, "The Garden Party," produced in Prague in 1963, was soon staged throughout Europe. "The Memorandum" and "The Increased Difficulty of Concentration," Havel's next two plays, both received Obie awards as best foreign play when presented by the New York Shakespeare Festival in the late 60's. More recently, "A Private View" and "Largo Desolato" played at the Public Theater in New York and at the Mark Taper Forum in Los Angeles. His work was banned in his own country from 1969 until just under a year ago. Havel's published writings in this country include "Letters to Olga: June 1979-September 1982," a book of letters to his wife from prison. He is a founding member of the human rights group Charter 77. Havel was elected president of Czechoslovakia on December 29, 1989.

HARVEY SEIFTER, Director, took on the artistic direction of the Magic last spring after serving for two years as the theatre's Managing Director. Prior to coming to the Magic, he spent seven years as Executive Director of New York's Theatre for the New City (TNC), where he produced more than 250 new plays, including 12 Obie Award-winning productions. He also directed and translated plays for TNC, commissioned more than 25 new works and negotiated the purchase of a new three-theatre complex in New York's East Village. Seifter has directed at the Provincetown Playhouse and at the Soho Rep, and has assisted Robert Wilson on his monumental "Civil Wars" at the Schauspielhaus in Cologne. His West Coast directing debut was last season's critically acclaimed production of Milan Kundera's "Jacques And His Master." Seifter has received fellowships from the Goethe Institute and the International Theatre Institute, served as a panelist for the National Endowment for the Arts, and ta seminars and master classes in t at New York University, Harva the Ecole Nationale de Paris.

Mary Anne Cook, the Magic Theatre's general manager, creates the playbill for each production with the help of marketing director Barbara Sansone, who uses a scanner and the page layout program Aldus Pagemaker. Advertisements, photographs taken during dress rehearsals and the Magic's logo are scanned into Pagemaker; the final electronic product is then presented on disk to a service bureau for printing. "We save money and lots of time this way," says Cook. "We can change the playbill until just a few days before opening night."

Dozens of elements go into each issue of the Magic Theatre playbill, including camera-ready ads as well as ads created by Cook and Sansone from logos supplied by advertisers. The ads are stored for future use, when they might be re-sized for a later playbill.

A Visual Education

Bringing photos and drawings into the classroom

From their earliest stages of development, children express themselves through pictures. Long before they can write or even articulate their thoughts verbally, they begin exploring and defining their world by drawing it. They learn visually as well, recognizing pictures of their favorite animals, for instance, long before they recognize the combination of letters that spells out the animal's name.

Their initial exposure to computers is also visually-based: kids love computer games and the small but growing number of visual, interactive computer learning programs. In schools, computers are touted as crucial tools in the process of teaching children to think creatively and to learn at their own pace.

Now, scanners allow young children to learn, and to explain what they've learned, by combining text and images. Reports and school projects, even those created by small children, can be illustrated with a student's drawings or with paintings and drawings from a favorite book. Maps can bring to life history or geography reports; charts, graphs and diagrams can help drive home key points in math and science projects. Along the way, using a scanner-computer combination to experiment with different sizes and shapes for the same image — by cropping, enlarging, reducing or rotating it — is a great way for kids to learn how images are edited and altered in movies, television programs and advertising, newspapers and magazines. That first-hand knowledge helps make them visually-savvy consumers in our multi-media world, in which they're exposed to countless images every day.

As they grow to become teenagers and young adults, students are increasingly asked to eschew images — despite the fact that they've grown up in a visual culture, and despite the preponderance of visual aids in their early classroom years. They're required to use words and only words to express themselves, and to understand concepts presented primarily through words. But high school and college students, too, can use image scanners to illustrate their reports and theses, to complement the written word. A research paper for a class in art history might feature a photograph of a painting, scanned in by a student, and cropped and enlarged to highlight a particular technique favored by the painter in question. A linguistics student might document a report on phonetics with a diagram of the mouth and speech organs, illustrating how various vowels are produced. A student of biology could scan into a report an original drawing of the structure of a cell, while a map of a dig site and a photo of a specific discovery would add another dimension to an archeology paper. Even more creatively, a teacher or professor might ask her students not only to write a critique of the Robert Frost poem "Stopping by the Woods on a Snowy Evening," but to gather a series of photographs that illustrate the imagery suggested by the poem — thus using photos to demonstrate the evocative power of language.

In the projects highlighted on the following pages, students and teachers use image scanners to expand the notion of knowledge beyond its traditional, word-based structure. The elementary school application of a scanner as a tool in writing biography is made even more powerful with its use as an autobiographical, almost therapeutic tool for disadvantaged students; a university student, meanwhile, finds that imagery lends added dimension to a graduate thesis.

TIPS for education use

❶ *As a vocabulary-building exercise, teachers can ask young children to match scanned images of objects or events with dictionary definitions of the same items.*

❷ *With a computer, a scanner and a laser printer, even elementary school students can produce a handsome school newspaper — for a fraction of the cost of producing school papers by traditional printing methods.*

❸ *To help students learn how to express themselves more clearly with words, teachers can scan a series of photos into the computer, print them out on the laser printer, and ask the students to write a story based on, and linking, the images.*

❹ *Students can be encouraged to include scanned images as part of their reports and research papers, and to create visual study aids.*

❺ *Before photocopying a photograph for classroom distribution, teachers should scan and laser-print the photograph first: the quality will be significantly higher.*

Expressive "Books" from Creative Young Authors

A decade after their introduction, personal computers are fixtures in thousands of American classrooms. Now, two innovative California teachers are using scanners to make learning more exciting than ever for a class of underprivileged 7th-graders. The students, many of whom have learning disabilities or speak little or no English, scan their drawings, family photos and mementos such as birth announcements into the computer. Using the Macintosh program known as HyperCard, they then create their autobiographies — as a way to learn more about themselves and to share with their classmates information about their ethnic backgrounds and cultures. "The kids also use the scanner to illustrate their reports," says teacher Jeanie Coltreaux. "They do the scanning themselves, and the computer allows them to go at their own pace." Adds teacher Karla Kelly, "The kids are so involved in the process of learning that they don't need to use their grades as a motivating factor."

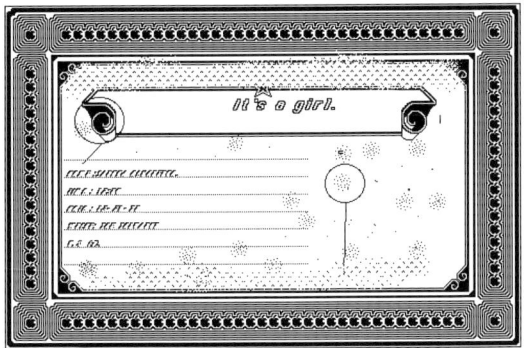
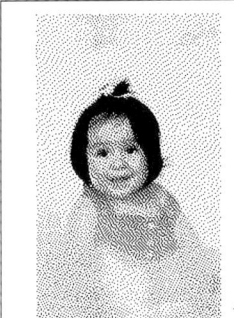

Seventh-graders at Davidson Middle School in San Rafael, California, use a scanner to illustrate their autobiographies in a special program sponsored by LucasFilms. "The kids are learning skills they never knew they had," says teacher Karla Kelley.

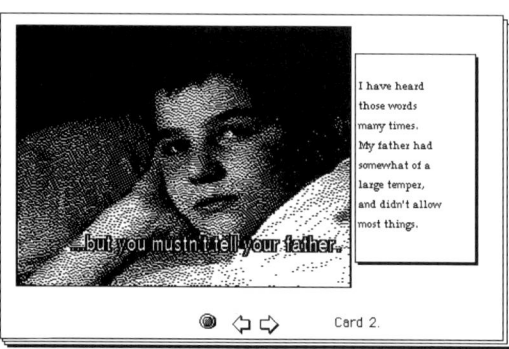

The students become so comfortable with their new communication skills that they're able to share sensitive information with their friends — things that are usually hard to put into words.

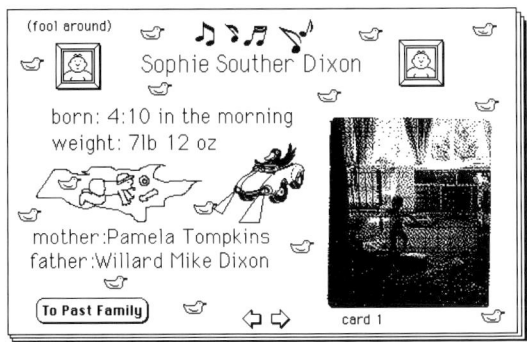
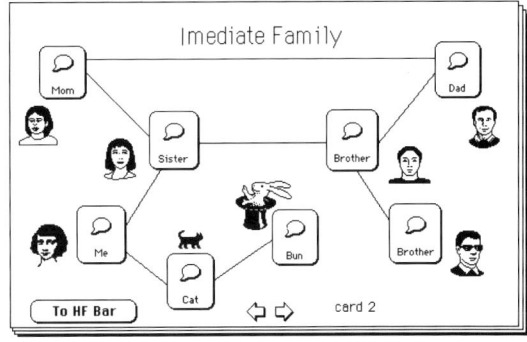
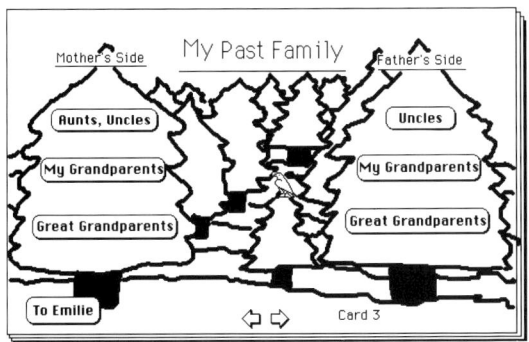
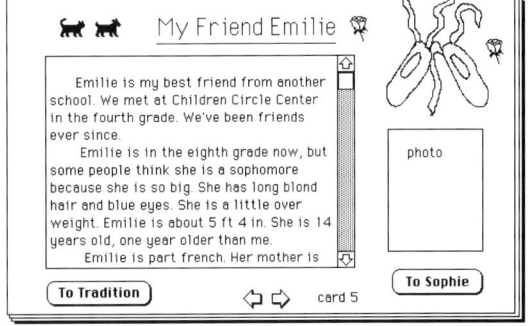

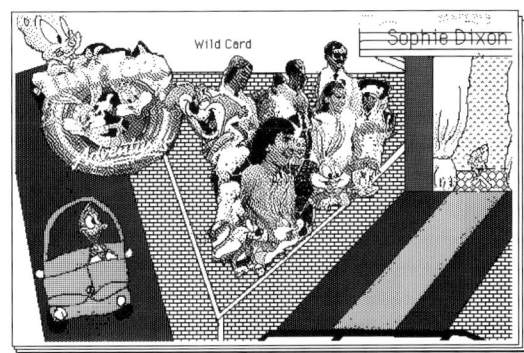

Working with a scanner also enables kids to discover artistic creativity that often isn't tapped by other school assignments. For the students at Davidson Middle School, writing autobiographies is fun as well as educational.

A Graduate Student Brings Her Thesis to Life

It's a paradox of education: teachers use images and words together — films, slides and generously illustrated books — to teach their students, while the tests and research papers turned in by the students are usually just words, words, words. That's no longer true in the age of the scanner: students such as Cynthia Liu at the University of California at Berkeley are finding that they can enhance their research papers and theses by including art and photographs. To illustrate her graduate thesis for a Master's degree in English, Liu scanned in traditional Chinese paintings. "The paintings were the inspiration for many of the authors my thesis discusses," says Liu, "and they help to place the reader in the proper historical epoch. It took almost no time to scan the paintings in, and they gave my thesis the look of a finished textbook."

Woman Warriors

The Story of Lady Wen-chi

Ts'ai Yen was the daughter of Ts'ai Yung, a scholar-statesman of the Han Dynasty. Also known as Lady Wen-chi, she was abducted around A.D. 195 and taken to inner Mongolia, where she was made the wife of a commander of the Hun army. Many of the poems based on her experiences explore her longing for her homeland and the dilemma she faced when a ransom mission found her years later. Should she return to her people and the land of her birth, leaving her husband and children behind, or should she stay with them in what would always remain to her an alien land?

According to *Eighteen Songs of a Nomad Flute: The Story of Lady Wen-chi*:
Wen-chi's decision to return to China became associated with some cherished ideas: the superiority of the Chinese civilization over the cultures beyond her borders; the irreconcilability of the different ways of life; the necessity for the individual to bear the burdens thrust upon one by fate, and above all, the Confucian concept of loyalty to one's ancestral family and state.

In many ways, the story of Ts'ai Yen typifies the challenges that faced many Chinese women during the first centuries A.D. They made great sacrifices to remain loyal to their homeland, where women traditionally assumed a subservient and often restricted role in society. These women, who refused to take advantage of the opportunities available in other lands, are called "woman warriors" by Hong Kingston in her novel of the same name. Does the designation "woman warriors" adequately reflect— or unnecessarily glorify— the significance of these women's choices and actions?

Scanned images, used by university student Cynthia Liu, enhance the informational content of her graduate thesis.

Designing Like a Pro
High-quality design on a desktop computer

Since the introduction of the first desktop publishing systems in the mid-1980s, desktop computers have held out the promise of making every user into a graphic designer — while, at the same time, helping professional graphic designers raise their art to even higher levels. And, over the past decade, computer technology has indeed revolutionized all forms of printed communication, from in-house corporate newsletters to high-powered design work in publishing and advertising. Page-layout and graphics software, large high-resolution monitors, powerful imagesetters and high-quality printers now allow people in a wide variety of professions to tackle projects that were once the exclusive domain of professional designers; professional designers, in turn, have found that computer technology allows them to experiment quickly with multiple options for each design, and to produce high-quality proofs of their work in a short period of time.

Perhaps the most frustrating element of the computer graphics revolution has been the software used with image scanners, most of which has been difficult to use and offers inconsistent quality. The same inadequate software that makes scanning time-consuming and sub-standard for professionals often makes scanning an impossibility for less experienced users. But the newest developments in scanning technology solve problems on several levels, and allow designers at all levels, from the novice to the celebrated, to work more efficiently, inexpensively and creatively.

In the past, many users found that using a scanner to produce a straight and perfectly placed image required several and sometimes dozens of scans, and also required a user's familiarity with the subtleties of scanner technology. Now, scanning has become a process more akin to using an office photocopy machine or fax machine. Professional designers can use scanners to work more quickly than ever, and less experienced users no longer have to familiarize themselves with concepts such as dots-per-inch, gray levels, image types and calibration in order to use a desktop scanner. The original image, once placed on the scanner's platten, is automatically straightened for output by the scanner — so that the image scans correctly on the very first try. Anyone who has ever used a desktop computer can quickly and easily make alterations in the scanned image's size, shape, rotation and cropping, and can be confident that the reproduced images will look like the original.

The quality of scanned images has also improved markedly, thanks to a technology known as adaptive calibration, which allows new scanning software to make precise calculations for the user, depending on the kind of image, the kind of paper and the kind of printer. The software can also automatically determine whether the image is line art or photograph and adjust accordingly, and will automatically adjust for brightness and contrast. Users can also enlarge or reduce an image without having to worry about "moires" — which is what designers call the wavy patterns that can rob a scanned or printed image of its sharpness.

Now, non-professionals can more closely approximate the levels of design sophistication that professional designers have come to expect. And from the artist at a leading design studio, to the designers who developed signs for one of the world's great museums, to the publisher of travel books who uses scanned images as final art, professional designers are finding that a scanner can make a major difference in how they approach their work. While not every person who uses a scanner shares their level of design experience, virtually everyone can learn from their examples.

TIPS for design use

❶ *Many designers make creative use of the fact that scanners are not limited to photographs and images; they can also scan fabric, objects and the like.*

❷ *Scanned images may be combined with commercially-available drawing software to good effect. Cartographers have found that they can create maps by scanning an aerial photograph into a computer and then tracing the outlines of the photograph. Fashion designers use image scanners to scan fabric into a desktop computer; they can then trace the old pattern, making any necessary alterations for the new season. (As always, copyright issues are important here: even if you change a copyrighted image, by scanning it into your computer and manipulating it with a drawing or image editing program, you still may be in violation of copyright law, depending on how extensively you alter the image.)*

❸ *Before outputting your photograph or drawing to an imagesetter, print it on a laser printer for proofing purposes.*

Artist Gives Top Marks to a Desktop Scanner's Crop Marks

Professional designers often approach new technologies a bit warily; their standards, after all, are excruciatingly high. But Kit Hinrichs of Pentagram Designs — the internationally known design firm that numbers among its clients The Nature Company and Royal Viking Lines — has become an enthusiastic user of a scanner. To select the exact crop of a photograph to be used in a catalog or annual report, for instance, Hinrichs and a computer artist scan in the photo, and experiment — quickly and easily on the screen of their computer — with cropping it several different ways. They then print out several versions of the photograph on their laser printer, so Hinrichs and his client can compare the various crops before settling on the final image. "Scanning the images and cropping them on screen saves time," says Hinrichs. "But most importantly, I can view numerous options instantaneously, which enables me to make smarter communications decisions for my clients."

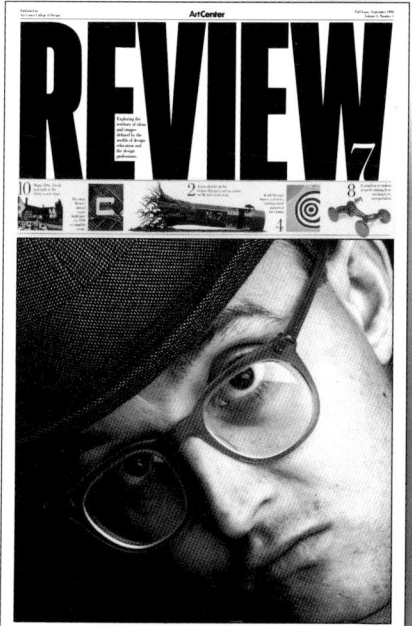

By using a scanner to experiment with a variety of crop positions, designer Kit Hinrichs can show a client — in this case, the Art Center of Design in Pasadena — several options before settling on a final cover design.

PHOTO: STEVEN HELLER

At the Louvre, the Sign of a Job Well Done

For the New York design firm of Carbone Smolan Associates, the opportunity to design the sign system for the Louvre museum in Paris was the chance of a lifetime. Six other finalists were in the running for the commission, meaning that Carbone Smolan's presentation of its concept would be critical. Company president Ken Carbone settled on the device of pictorial icons to anchor Carbone Smolan's design. To present the design to the jury in charge of choosing a design firm, Carbone scanned images of four recognizable Louvre works of art directly into the drawings for sign prototypes. "Visual aids are the best way to explain a visual concept," says partner Leslie Smolan, "so rather than explain what we proposed to do, we decided to show what we proposed to do." Adds Carbone, "Scanned images give proposals a more finished look, which in turn gives clients an immediate sense of the level of quality that will be achieved in the final product."

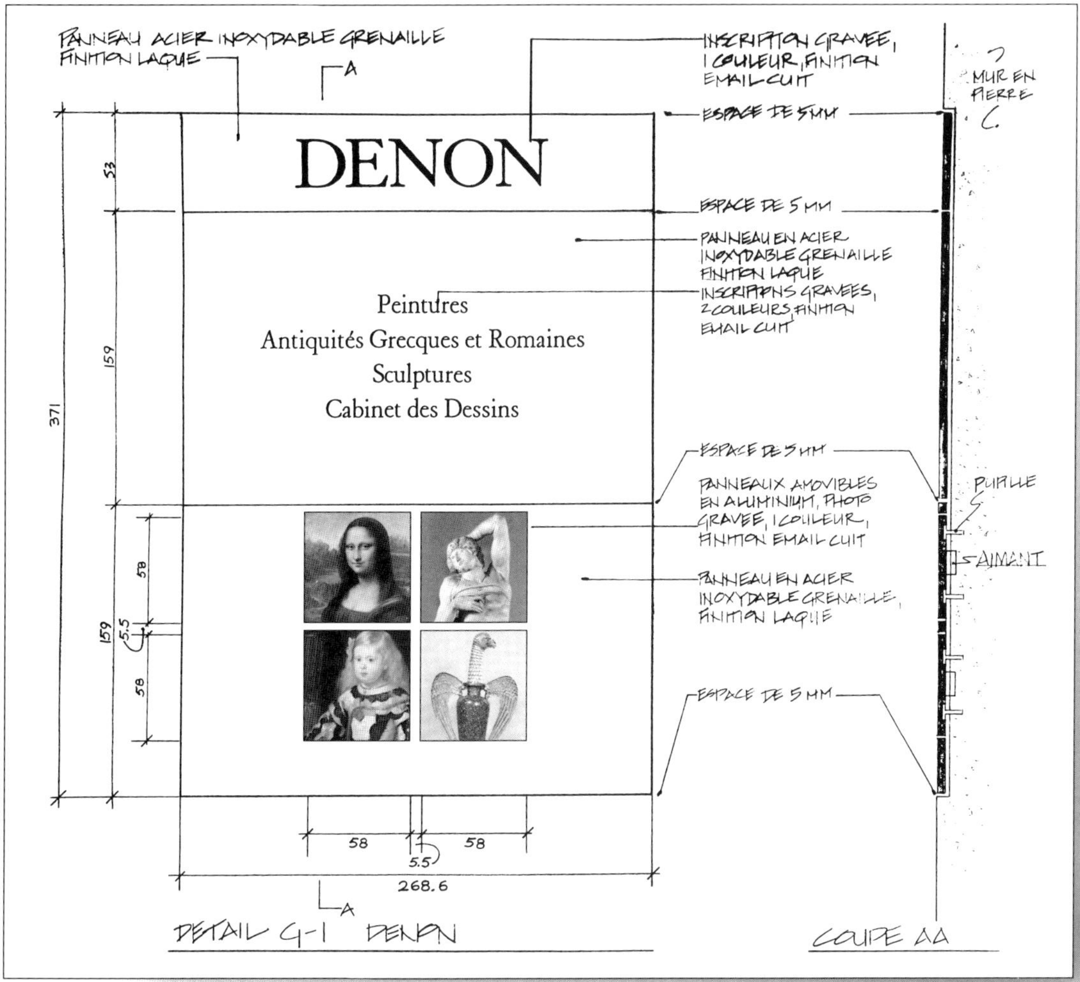

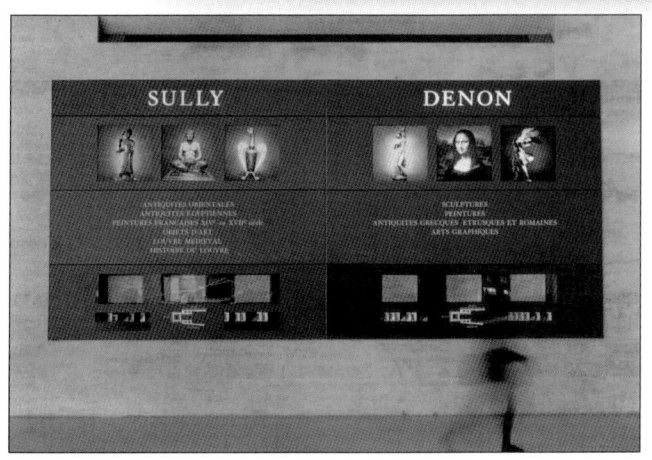

Images scanned directly into the Carbone Smolan Associates proposal for the signage at the Louvre gave museum administrators the opportunity to see exactly what the designers were proposing. Left: The new signs at the Louvre, designed by Carbone Smolan.

Drawing on the Latest Technology to Illustrate Travel Books

Much of the appeal of the famed "Access" travel books comes from their unique design and creative use of drawings. The books have relied on computer publishing technology since the series debuted with "LA Access" in 1980; now, Access editors find that they can push their creativity even further with the speed, quality and flexibility offered by the use of a scanner. With the scanner, explains Access founder Richard Saul Wurman, publishers can insert line art — such as original drawings or other, non-original drawings for which permission has been obtained — directly into a computer file, in the proper position in relation to the text. The entire file can then be printed directly, without designers at a print shop or service bureau having to place the art. "Before the scanner," says Wurman, "the only other way to place line art directly into a file was to draw it on the computer with a commercial drawing package."

A page from the book "SF Access", with line art scanned directly into the desired position in a computer file and printed right onto the page.

Civic Center

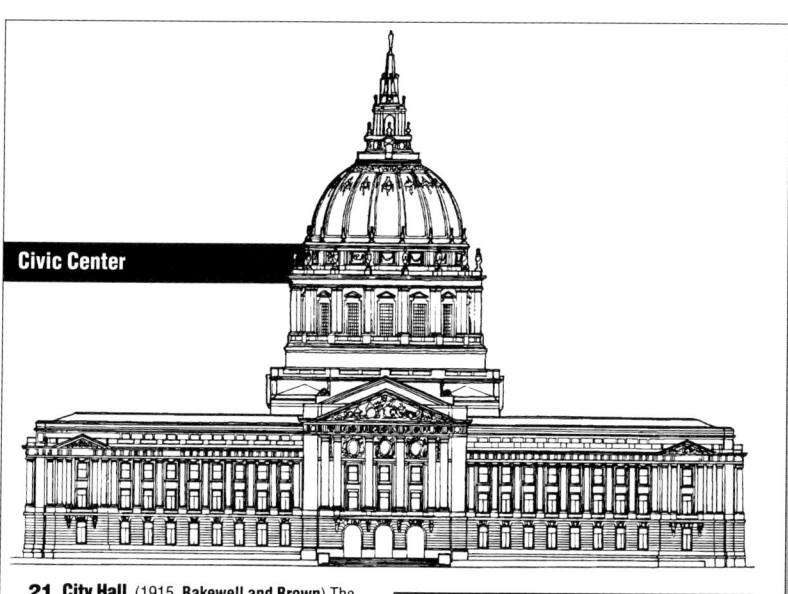

21 City Hall (1915, **Bakewell and Brown**) The focal point of the **Civic Center Complex** is City Hall, a magnificent symbol of government with its huge dome (modeled after St. Peter's in Rome), Baroque stairs, and echoing marble-clad corridors. The plaza in front of City Hall is disappointing by comparison; it's at its best when used for civic functions. Embarrassingly full of litter, in recent years it has become a magnet for vagrants. ♦ Van Ness Ave at McAllister St

Drawing courtesy Carlos Diniz

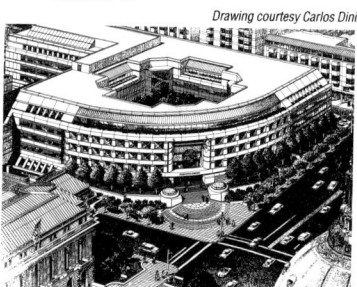

22 State Office Building (Edmund G. Brown Building) (1986, **Skidmore, Owings & Merrill**) The design complements the Louise M. Davies Symphony Hall by facing diagonally toward City Hall and completing the Beaux-Arts composition of civic buildings along this stretch of Van Ness Avenue. Clad in white precast concrete, it has a large seal of the State of California above the entry to the courtyard, which is disappointingly institutional in scale. ♦ Van Ness Ave at McAllister St

One of San Francisco's most important industries is **tourism**. Every year visitors outnumber the resident population by 5 times, with almost a third of these 3 million people coming for conventions.

12

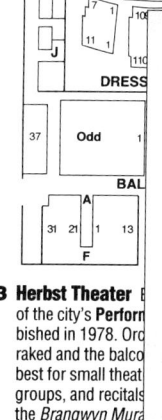

23 Herbst Theater of the city's **Perform** bished in 1978. Ord raked and the balco best for small theat groups, and recitals the *Brangwyn Mura* Panama-Pacific Exp 928. 401 Van Ness

Free **walking** tours of San F are given throughout the week by **City Guides and Friends of the Public Library**. Information: 558.3981.

Dancer **Isadora Duncan** opened a dancing school at Van Ness Ave and Sutter St.

Restaurants/Clubs: Red **Hotels:** Blue
Shops/Parks: Green **Sights/Culture:** Black

Pictures on File

Storing important images for multiple use

Many business people find that the most attractive element of a computer is not its ability to create text and information, but rather its capacity to store text and information. And while computers have specialized in storing text and other written data for years, only in very recent years, with the advent of scanners, have they been widely used to store digitized graphic information, such as photographs and line art.

Storing and sorting images electronically represents an important technological leap forward for business owners in dozens of fields. Architects can store plans as well as photos of finished work. Catalog designers can store photos of merchandise, along with text descriptions of each item, to mix and match in future editions of the catalog or to create custom catalogs and order sheets. Artists can store photos of their work. Photo agencies, which normally operate with what seem like miles and miles of file cabinets, can store photographs and use key word searches to quickly and easily find desired photos. Factory managers can store floor plans and machinery diagrams. Insurance agents can store photos of their clients' possessions; lawyers can store photographs of key exhibits. Around the house, a family can store treasured photos and children's drawings. With a scanner and a computer, any image that can be stored on paper in a file cabinet or photo album can be stored digitially.

The problem, however, is finding electronic storage space. Images gobble up far more space in a computer's memory than do text documents, to the extent that less than three dozen images could fill up an entire 40-megabyte hard drive. Many scanned images are too large to fit on a single floppy disk, making it problematic for the disk to be effectively transferred to another computer — at, say, a service bureau. Hard drives with massive storage capacities, along with alternate storage devices, offer some relief from the problem. And compression technology, which shrinks large files to much smaller sizes for storage purposes, has developed in recent years as another alternative, and is ever-improving.

But a large part of the storage problem is, in reality, a scanning problem. Because scanning software has been so imprecise in helping a user determine what scaling parameters to use in order to obtain a clear and crisp paper version of a scanned image, users are instructed to play it safe by scanning at the highest possible resolution, or dots-per-inch (dpi). High resolution means large image size, which also means storing the image at the largest possible file size.

New developments in scanning software tackle the storage space problem, thanks to the highest-quality and most precise scaling technology ever used with microcomputers. Where scanning software packages formerly required an image to be scanned at 300 dpi, it's now possible to scan the same image at 150 dpi — with comparable or superior results. In terms of file size, an image scanned at 150 dpi takes up only one-fourth as much space, even before being compressed, as an image scanned at 300 dpi. By applying the most modern compression techniques, the size of the stored file can be reduced even further — bringing image archiving and storage truly into reach for individuals and businesses who have been reluctant, in the past, to tie up the large amounts of storage space required by scanned images.

As the photographer and the casting agent profiled in the following section demonstrate, images can now be stored and accessed in much the same way that words are stored and accessed. And, as the owners of a prop and scenery business explain, image storage and retrieval is a highly effective inventory and selling tool.

TIPS for image archiving

❶ *Scanned photos of a firm's employees can be stored in a computer. The photos can be used in memos introducing a new hire or announcing a promotion, or in an employee handbook or personnel file. The photos can also be used to distribute by electronic mail, before an important meeting, images of all the key participants.*

❷ *A customized inventory of images might include photos of satisfied customers using your company's product — with their permission, of course. The photos can be used in promotional material, along with any photos of your company's representatives or clients that appear in newspapers, magazines and other publications.*

❸ *If you or your employees travel a lot, you can store in your computer a scanned copy of maps and directions to specific locations.*

❹ *By using the most up-to-date compression technology, you increase your computer's storage capacity.*

Tracking Inventory a Snap for Photo Agency

Just as a quality word-processor is indispensable for a writer, scanners are becoming indispensable for photographers. "Scanners have finally gotten to the point where what you see as your original image is what you get on the laser printer," says Doug Menuez, whose agency, REPORTÁGE, specializes in magazine and corporate work. "The output is terrific, and the software is becoming easy to use." Menuez uses his scanner to manage the flow of photos in and out of his agency: "I no longer have to send out dozens of original slides and wait months for them to be returned," he says, "because I can send high-quality scanned reproductions, and only have to send out the originals of the exact slides the art director chooses to use." He also uses the scanner to share his work with potential clients via regular promotional memos — letters that combine text with scanned reproductions of REPORTÁGE's most recent sales.

CLEMENT MOK DESIGNS – GREG
ROLL: 2454   FRAME: 5

CLEMENT MOK DESIGNS – KEVIN
ROLL: 2454   FRAME: 12

CLEMENT MOK DESIGNS – DALE & KEVIN
ROLL: 2455   FRAME: 14

CLEMENT MOK DESIGNS – DALE & GREG
ROLL: 2455   FRAME: 6

CLEMENT MOK DESIGNS – DORIS & MARK
ROLL: 2456   FRAME: 22

CLEMENT MOK DESIGNS – BILL
ROLL: 2456   FRAME: 24

CLEMENT MOK DESIGNS – KEVIN
ROLL: 2456   FRAME: 8

EDUCATION SERIES #3
ROLL: 1724   FRAME: 6

EDUCATION SERIES #3
ROLL: 1743   FRAME: 24

EDUCATION SERIES #3
ROLL: 1799   FRAME: 23

EDUCATION SERIES #7
ROLL: 1820   FRAME: 2

Instead of sending out dozens of contact sheets and original slides (below) — only a fraction of which may precisely suit to a client's needs — photographer Doug Menuez scans into his computer images from various takes, using the print-out to provide the client with a customized and specific contact sheet (left).

Once a client has chosen the desired originals, Menuez no longer has to send along with them large and muddy photocopies (below) as a record of what was sent. Now, he appends to the originals a delivery memo with scanned reproductions of the original frames, so that Menuez and the client both have a record of which photographs were received.

REPORTAGE

DELIVERY MEMO

Joanne Hoffman
✉ Macworld Magazine
501 Second Street
San Francisco, CA 94107
415/978-3326

Date: March 14, 1991
Return by: March 28, 1991

Re: Computer Graphics

Memo: 1/4 page - interior usage

Count	Description	Format:
4	Apple Computer	35mm
6	Clement Mok Design	35mm
10	Total	

TERMS AND CONDITIONS:

All photos COPYRIGHT 1990 DOUG MENUEZ/REPORTAGE. All photos published should carry this notice. All photographic material, transparencies, negatives or prints must be returned in good condition within 30 days, unless covered by specific agreement in writing. Return must be made by registered mail or messenger, fully insured, return receipt requested. All material not returned within 30 days will be billed $5.00 per week per image unless otherwise authorized.   LOST ORIGINAL NEGATIVES OR TRANSPARENCIES WILL BE BILLED AT $1500.00 EACH. Rights granted are for one-time editorial use only. This does not include reuse in any form of promotion or advertising. Rights granted by this invoice become effective with signature below.

Acknowledged and accepted _____ Date _____

Publication/Company _____

Please sign and return this form upon receipt of material

REPORTAGE

Liz Faulkner
Against All Odds Productions
1103-04 57 Street, Suite 6N
New York, NY 10024

5-2-91

Dear Liz,

I'm writing to let you know what we at REPORTÁGE have been up to lately. You may have seen our two recent covers of Fortune; I've reproduced them here for you to have a look.

As I think you know, we specialize in bringing the power of photojournalism into the world of corporate communications. We're strong believers that dynamic, candid photojournalism gives added zest to brochures, annual reports and advertising campaigns. And we like to work closely with our clients, shooting over a long period of time to build a library of images that can be used at various times and in various ways.

I hope you'll consider us next time Against All Odds Productions wants to capture the emotion, quiet interactions and sudden surprises of everyday working life.

Sincerely,

Dm

Doug Menuez

P.O. Box 2284   Sausalito, CA 94966-2284   415/332.8154

By combining text with images, Menuez can keep prospective clients abreast of his work.

Prop Keepers Communicate with Clients without Making a Scene

Gray McKee and Josh Koral, the owners of ACME Scenery Company, pride themselves in their eclectic, extensive and ever-expanding collection of props for theatre and film. Such a large and varied collection, however, is hard to keep track of — and just as hard to market. "We want to show exactly what our items look like, and must determine whether an item is in stock," says McKee. "And we need an efficient way to update our list to include new items." With a scanner, McKee and Koral can digitally store photos of each of their 5,000 items, and they can annotate a photo to indicate whether the item is in stock. They can create a general catalog, complete with images and descriptions, as well as custom mini-catalogs for clients who call with specific requests. "For the first time," says McKee, "we'll know exactly where everything is, and customers can see exactly what we have."

A scanned-in logo can be used on the cover of Acme Scenery Company's prop catalog; inside, photographs can accompany descriptions of the items stored in the company's 600-square-foot warehouse. Partners Gray McKee and Josh Koral can also create custom catalogs — complete with scanned-in photos that have been stored digitally and categorized for easy retrieval — for clients who call with specific requests.

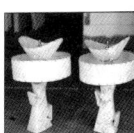

·A·L·O·G   ·2·

lls
. If we can hold it up,
u. 3 feet high, 7 feet
ot thick. Looks like the
. Breaks easily for fight
rge quantity available
be painted. Also in stock
91: stucco, brick, raised
el, plain and fabric flats
in assorted weights and

8. A matched set, arty and
ary. 3 feet high. Available
ntals come with bulbs
ady to plug in and play.
ck as of 5-1-91: table
er floor lamps—period
mporary.

Pedestals
Item #172. A marvel in marble. Four feet high, two feet wide. Twelve available now, with custom decoration available. Also in stock as of 5-1-91: dozens of pedestals in a variety of sizes and finishes.

Set Decorations
Item #499. A perfect prop to give your set a realistic, lived-in look. Set of one dozen available now. Also in stock as of 5-1-91: statues, ashtrays, small pictures, fake and real books, all sizes and styles.

For price/ordering
Information, see page 31.
To arrange for delivery and
pickup, call 415-468-2262.

Gray McKee and Josh Koral used to track their inventory with a typed list of items, which was sent to potential customers on request. McKee and Koral were available to describe items over the phone, but clients usually found it impossible to settle on an order without visiting the ACME warehouse to see for themselves.

A Casting Agent Sends Out Just the Fax, Ma'am

Nancy Hayes, owner and operator of a casting agency for films and commercials, relies on photographs as key selling tools. And faxed photographs of her actors arrive at their destinations in clean and readable form, now that she uses a scanner. Instead of photocopying an actor's photograph and faxing the photocopy to a prospective client, Hayes prints — on her office laser printer — a scanned photograph directly from her computer. She then faxes the prospective client a printout of the scanned photograph. The computer-scanner combination also allows Hayes to digitally store and categorize, by physical type, the photographs of her actors. Until recently, she stapled a snapshot of each actor to a casting sheet listing the actor's skills and vital statistics; now, she scans the photograph directly into the actor's casting sheet in the computer, crops and enlarges the photo to highlight the face, and prints the casting sheet as one piece.

In a system that is cleaner and more efficient than the stapling of a snapshot to a casting form (below), Nancy Hayes scans an actor's photograph directly into the casting form, after enlarging and then cropping the photo to create a head shot.

EXTRA FORM

Nancy Hayes Casting

Date _____ Social Security # _____
Name _____
Address _____
City _____ State _____ Zip _____
Birthdate _____ Age _____
If under 18, do you have a current work permit? _____

Phone _____ Machine _____
Height _____ Weight _____ Hair Color _____ Eye Color _____
Do you have an agent? _____

Men/Boys Sizes _____ **Women/Girls Sizes** _____
Shirt _____ Blouse _____
Jacket _____ Dress _____
Slacks _____ Pants _____
Shoes _____ Shoes _____
Hat _____ Hat _____
Glove _____ Glove _____

Do you own a tuxedo? _____ Business suit? _____
Do you own evening clothes? _____

Special skills (include singing, dancing, sports, foreign languages)
_____
_____
_____

Do you own a car? _____
Make _____ Model _____ Year _____ Color _____
Motorcycle? _____

Union Affiliations Please check:
☐ SAG   ☐ SEG   ☐ AFTRA   ☐ AEA

Have you had experience as an extra in film or television? _____

From the original photograph (center), Nancy Hayes laser-prints a scanned version (lower left) and faxes it to a prospective client (lower right). Before she began using a scanner, she would photocopy the photograph (upper left) — leaving the client with a muddy fax (upper right).

Printing: the Final Step
Getting your work out on paper

For both the business user and the home user, scanning images into a desktop computer is not the final step in the creation of a finished document that includes both text and images. In fact, even once the final document is composed in a page-layout program and saved in the computer, a key issue remains: how, and where, to print the document.

Most business users who want crisp and attractive documents — but don't have the same requirements of professional publishers or design firms — find that the choice is an easy one: a good laser printer offers high-quality output, which in turn can be reproduced on high-quality photocopy machines. Laser printers are ideal for most business uses, including sales proposals and business letters with scanned-in photographs, as well as internal documents and forms. Permission to use the scanned images, of course, should be obtained before printing begins; many desktop publishers have found a way around copyright problems by using copyright-free "clip art" images to dress up their documents. (A significant collection of copyright-free line drawings and photographs is available in the Dover series of books, published by Dover Publications of Mineola, New York. Copyright-free art is also available on compact disc, for computer users whose hardware includes a CD-ROM player. The Bureau of Electronic Publishing of Parsippany, New York, publishes a catalog of clip art information for owners of CD-ROM drives for both Macintosh and IBM-compatible computers.)

Jobs that require high quality and high volume are best handled by offset printing, a technique that print shops began using in the 1950s. With the combination of a desktop computer and scanner, users can now deliver their documents to a print shop in various stages of readiness for the printing process. And regardless of what form your computer-generated document is

Laser printers are ideal for most business uses, including proposals and letters with scanned-in photographs.

By adding a high-quality, easy-to-use scanner to the desktop publishing equipment, preparing a text-and-image document can be handled by one person at home or at the office.

in by the time you take it to a print shop, it's virtually a certainty that you will have leapfrogged at least one step in the traditional printing process.

The key to saving time and money in the preparation of a document for offset printing is the use of an imagesetter — a piece of hardware often referred to by the name of a leading manufacturer of imagesetters, Linotronic. Many publishers and professional designers own or lease imagesetters; others get imagesetter output by taking their document files on computer disks (or sending the files by modem over the telephone lines) to imagesetter-equipped service bureaus — which are found in increasing numbers in business centers and commercial districts across the country — or other so-called "pre-press" establishments. As computer-aided publishing continues to grow, many print shops now feature imagesetting services, just as many service bureaus now feature offset printing capabilities.

The imagesetter, which typically prints documents at either 1,270 or 2,540 dots-per-inch (dpi) — as opposed to the 300-dpi output afforded by a laser printer — can create either paper or film versions of a document. At a print shop, either the paper or film versions are then transferred to a printing plate, made of metal or paper, from which the offset press actually prints. You can, if you prefer, also use a high-quality photocopy machine to copy the paper version of your document after it has been through the imagesetter; your final product will be sharper than if it had been printed on a laser printer and then photocopied, but not as sharp as it would be if printed offset from imagesetter paper or film.

Until the use of scanners and imagesetters became widespread, a client typically presented a print shop with typeset text, each page laid out and waxed onto a stiff piece of paper called a "flat" or a "board." On each board, the client had blocked out, with red acetate paper, the space to be occupied by black-and-white photographs or line art; the print shop then made "camera-ready" duplicates of the images supplied by the client — following the client's instructions for cropping, enlarging, reducing or rotating each image. (This duplication, which was necessary in order to bring out all the contrast and shading in the images, is typically achieved through a process known as "halftoning," or "creating a halftone screen.") The print shop then prepared film negatives of each page, complete with camera-ready images "stripped in" to replace the red acetate blocks in the positions indicated by the client, and transferred the film negatives to metal plates in order for the printing process to begin.

Such a process, besides being unwieldy, deprived the client of being able to see exactly what each page might look like with the images incorporated. It also deprived the client of being able to experiment with various sizes and various crops for each image, unless the client was willing to spend money ahead of time for camera-ready duplications — or "photostats" — of the images in a variety of sizes.

Now, with a scanner and ofoto™ scanning software — featuring the precise calibration, straightening of images, file size reduction, ease of operation, and many uses described throughout "Worth a Thousand Words" — clients can prepare complete pages within a computer file, with the images scanned in to their place on each page. If a client opts for paper output from an imagesetter, a print shop can then take the camera-ready page that the imagesetter makes from the computer file — a page that substitutes for the "board" or "flat" used in traditional methods — and create the film negatives, and then the metal plate, from which to print. The client can jump ahead a step by receiving film negatives directly from the imagesetter — a process known as "going directly to film." (When going directly to film with a computer-generated document, it's a good idea to make a laser-printed comp first, in order to proof the pages.)

Choosing between various print options is the final step in computer-aided publishing. But regardless of how a file is printed, the use of a scanner has given creative people in all walks of life greater control over their work. At the office or at home, projects can be completed more quickly, and more options can be explored before a final design is chosen.

TIPS on copyright law

❶ *Clip art is not the only option for copyright-free art: many stock photography agencies sell ready-to-use photographs.*

❷ *When you ask for permission to use a copyrighted image, state your intentions in detail — and adhere to the terms of your usage agreement.*

❸ *Generally speaking, the copyright for an image stays in effect for 50 years after the death of the work's creator, after which the image falls into the public domain. At that point, it's fair game, and anyone can use it. Be careful, though: many famous works of art are owned by museums that, as part of their ownership agreement, retain exclusive rights to the image's reproduction.*

❹ *Further information on copyright law can be obtained from the Copyright Office at the Library of Congress, Washington, DC, 20559. Telephone (202) 707-9100.*

Comps and Scans at a Neighborhood Service Bureau

For Diane Burns, president of TechArt design studio and service bureau, the scanner has become an indispensible business tool. Not only does TechArt have the ability to scan images for individual customers who don't have scanners of their own, but the company also uses the scanner in designing materials, start to finish, for commercial clients. "We use the scanner to produce halftone images," says Burns. "The client can then review the draft, or comp, and give us feedback about any necessary changes — which we can then go back and make simply by manipulating the scanned image."

Before the advent of high-quality and easy-to-use scanners, Burns found herself producing comps for her clients and then having a print shop strip in halftones. "Preparing comps and final layouts is much easier and less expensive with a desktop scanner," says Burns. "And the quality is so good that clients can see exactly what they're getting."